Reptile Adventure

Alligators

MARK WEAKLAND

Black Rabbit Books

Bolt is published by Black Rabbit Books
P.O. Box 227, Mankato, Minnesota, 56002
www.blackrabbitbooks.com
Copyright © 2024 Black Rabbit Books

Alissa Thielges, editor; Michael Sellner, designer
and photo researcher

All rights reserved. No part of this book may be
reproduced, stored in a retrieval system or transmitted in
any form or by any means, electronic, mechanical, photocopying,
recording, or otherwise, without written permission from the publisher.

Library of Congress Cataloging-in-Publication Data
Names: Weakland, Mark, author.
Title: Alligators / Mark Weakland.
Description: Mankato, Minnesota: Black Rabbit Books, [2024] |
Series: Bolt | Includes bibliographical references and index. |
Audience: Ages 8-12 | Audience: Grades 4-6
Identifiers: LCCN 2023001219 (print) | LCCN 2023001220 (ebook) |
ISBN 9781623106409 (library binding) | ISBN 9781623106461 (ebook)
Subjects: LCSH: Alligators—Juvenile literature.
Classification: LCC QL666.C925 W378 2024 (print) | LCC QL666.C925
(ebook) | DDC 597.98/4—dc23/eng/20230501
LC record available at https://lccn.loc.gov/2023001219
LC ebook record available at https://lccn.loc.gov/2023001220

Printed in China

Image Credits

Alamy/Colin Pickett 4–5, Gregory Wrona 17 (b), image-BROKER 6, Itsik Marom 3, 32, Mark Andrew Thomas 18–19, Natural History Archive 22, Nature Picture Library 28 (egg), Xinhua 28 (baby); Dreamstime Isselee 6–7, 28–29 (teeth), Le Thuy Do 24–25, Michalna-partowicz 10–11, Pindiyath100 20–21; Getty/Newspix 25, PJPhoto69 14 (tail), Tim Chapman 26, Wirestock 23; Shutterstock/Brian Lasenby 15, Chuck Wagner 14 (feet), Heather LaVelle 24, Heiko Kiera cover, Irina Kryvets 12–13, John Cawthron 17 (t), Patrick Rolands 1, reptiles4all 8–9, 31, Sergii Chernov 17 (m)

Contents

CHAPTER 1
Gone in a Gulp. 4

CHAPTER 2
Where They Live. 10

CHAPTER 3
What They Eat. 16

CHAPTER 4
Alligator Family 20

Other Resources. 30

CHAPTER 1

Gone in a Gulp

A fat frog sits on a riverbank. Suddenly, the river water in front of the frog explodes upward. An alligator **lunges** with its tooth-filled mouth gaping. There's a snap and a violent gulp. The frog is gone forever.

How Big Is an American Alligator?

WEIGHT UP TO 1,000 POUNDS (453.6 KG)

Big and Hungry

Alligators are big reptiles with big **appetites**. Only two types live on the Earth. One is the rare Chinese alligator. The other is the American alligator.

The American alligator is huge. It can weigh up to 1,000 pounds (454 kilograms). It can reach 15 feet (4.6 meters) in length. But its Chinese relative is much smaller. This one may reach 7 feet (2.1 m) long and weigh 100 pounds (45 kg).

LENGTH
UP TO 15
FEET
(4.6 m)

PARTS OF AN ALLIGATOR

SHORT LEGS

THICK, POWERFUL TAIL

CLAWS

CHAPTER 2

Where They Live

American alligators live in the southeastern United States. Many live in Florida. Chinese alligators live in a small part of the Yangtze River in China. All alligators live near rivers, swamps, and lakes. They are mostly freshwater animals.

The word "alligator" comes from the Spanish words *el lagarto*. They mean "the lizard."

Features Used to Swim

tail helps steer

webbed feet help swim

large head... pushes through underbrush

Adaptations

Alligators are well suited for wetlands. Their feet are webbed for swimming. Their tough scales, called scutes, protect them from the sun. It also helps hide them. Floating gators look like floating logs.

Alligator eyes, ears, and nostrils sit on the top of their head. This placement lets them see and breathe even when they are mostly underwater.

CHAPTER 3

What They

Alligators eat meat. They grab and gulp fish and frogs. They gobble down birds. Small mammals that come to the water's edge are also food. Rodents often get eaten. So do bigger mammals, such as deer.

Alligators bite down with a powerful snap. But the muscles that open their jaws are weak. One strong student could hold an alligator's jaws shut. But don't try it!

Big Bites

An alligator often swallows small animals whole. But if prey is too big for one bite, a gator has a trick. It grabs the large animal in its strong jaws. Then it spins in the water. The spinning tears off bite-sized bits.

CHAPTER 4

Alligator Family

Alligators are **cold-blooded**. They spend most of their days basking in the sun to stay warm. At night, they often hunt.

Male alligators, called bulls, also protect their **territory**. They **roam** far and wide. Females stay in smaller areas.

HEIGHT
2 to 3 feet high
(0.6 to 0.9 m)

WIDTH
7 to 10 feet across
(2.1 to 3 m)

Having Babies

Alligators come together to **mate**. For American alligators, mating happens in the spring. The males roar to attract females. The roaring also scares away other males. Chinese alligators often mate in the summer. Male and female Chinese alligators bellow and roar to each other. All female alligators build nests after mating.

An **American Alligator Nest**
made of mud, sticks, grass and other plants

Growing Gators

An American alligator might lay 30 to 70 eggs. Meanwhile, Chinese alligators often have 10 to 40 eggs. Once laid, the females cover the eggs with plants and sticks. For two months, the babies grow in the eggs.

Once the babies hatch, female alligators try to protect their offspring. But many babies don't grow to be adults. Hawks and snakes eat them. So do meat-eating mammals and other alligators.

Babies inside eggs make high-pitched noises. They do this when they are almost ready to hatch. The tiny squeals let the mother know it's time to uncover the eggs.

Threatened

Adult alligators don't have to worry about many animals hunting them. But people are a danger to alligators of all sizes. People destroy alligator **habitats**. They **dam** rivers. They drain swamps and lakes. People need to respect these giant reptiles.

Scientists think there are less than 120 Chinese alligators left in the wild.

By the Numbers

30 GRAMS (1.1 ounces) — weight of a Chinese alligator hatchling

3 inches (7.6 centimeters) — length of an American alligator egg

35 to 50 years — life span of a wild American alligator

3,000 NUMBER OF TEETH ONE AMERICAN ALLIGATOR CAN GROW AND REPLACE

ABOUT 85 MILLION YEARS how long alligators have been on Earth

GLOSSARY

appetite (AH-puh-tyt)—the desire to eat

cold-blooded (kohld-BLUHD-id)—having body temperatures that change according to the temperature of their surroundings

dam (DAM)—to put up a barrier that prevents the flow of water

habitat (HAB-uh-tat)—the place where a plant or animal grows or lives

lunge (LUHNJ)—a sudden forward rush or reach

mate (MAYT)—to join together to produce young

roam (ROHM)—to go from place to place with no fixed purpose or direction

territory (TER-uh-tor-ee)—an area that is occupied and defended by an animal or groups of animals

LEARN MORE

BOOKS

Feldman, Thea. *Alligators and Crocodiles Can't Chew! And Other Amazing Facts.* Super Facts for Super Kids. New York: Simon Spotlight, 2021.

Martin, Emmett. *Alligators from Head to Tail.* Animals from Head to Tail. New York: Gareth Stevens Publishing, 2021.

Sommer, Nathan. *American Alligator vs. Wild Boar.* Animal Battles. Minneapolis: Bellwether Media, 2023.

WEBSITES

American Alligator
kids.nationalgeographic.com/animals/reptiles/facts/american-alligator

American Alligator
sdzwildlifeexplorers.org/animals/american-alligator

Chinese Alligator
nationalzoo.si.edu/animals/chinese-alligator

INDEX

A
American alligators, 6–7, 10, 12, 23, 24, 28, 29

B
babies, 23, 24, 25

C
Chinese alligators, 7, 10, 13, 23, 24, 27, 28

E
eggs, 24, 25, 28

H
habitats, 10, 12–13, 27

hunting, 4, 19, 20

J
jaws, 17, 19

M
mating, 23

P
prey, 4, 16, 19, 24

S
senses, 15, 28

sizes, 6–7, 27